Kedar Nath tragedy: A shriek of abrupt Himalayan Environment

Dr. Hemant Pathak

DEDICATION

Dedicated to Shri Sainath Maharaj the all omnipotent of world the most merciful.

CONTENTS

Foreword

Kedar Nath tragedy: A shriek of abrupt Himalayan Environment; provides a unique insight into the problems our Nation (India) faces in terms of environmental calamity in Uttrakhand, India, and what is the causes behind it. This is the first book expressed their focus on Kedar Nath tragedy with the multidimensional approach.

This book made of 07 years consistently research on environmental issues, makes it ideal source for students, teachers, industrialist, environmental experts and economists.

This book provides an essential guide to environmental project managers working in Uttrakhand India, it offers: various aspects of problems; on the challenges and experiences in present scenario.

Simply explained, Kedar Nath tragedy: A shriek of abrupt Himalayan Environment is an important book for all who wish to make a difference in how to conserve and eco-management our Uttrakhand (The land of God).

Dr. Hemant Pathak

M.Sc. (Gold medalist), Ph. D.

Assistant Professor of Engineering Chemistry

Indira Gandhi Govt. Engineering College,

Sagar, MP, India

Glossary

Abatement- Reducing the degree of intensity of , or eliminating, pollution.

Acid- A corrosive solution with a Ph of less than 7

Act- in the legislative sense, a <u>bill</u> or measure passed by both houses of Congress; a law.

Alternative energy - energy that is not popularly used and is usually environmentally sound, such as solar or wind energy (as opposed to <u>fossil fuels</u>).

Amendment - a change or addition to an existing law or rule.

Aquifer - underground source of water.

Atmosphere- The mass of air surrounding the Earth.

Biodiversity-A short form of the phrase 'biological diversity', which means the variety of life on this planet and how it interacts within habitats and ecosystems. Biodiversity covers all plants, animals and micro-organisms on land and in water. See also ecosystem, habitat and organism.

Bioenergy- All types of energy derived from biomass, including biofuels.

Biofuels- Liquid transport fuels made from biomass.

Biomass- A source of fuel made from living and recently-dead plant materials such as wood, leaves and the biodegradable part of industrial and municipal waste.

Biosphere- The portion of Earth and its atmosphere that can support life

Biosphere Reserve - Biosphere Reserves are vital centers of biodiversity where research and monitoring activities are conducted, with the participation of local communities, to protect and preserve healthy natural systems threatened by development. The global system currently includes 324 reserves in 83 countries.

Biotic - of or relating to life.

Climate- The pattern of weather in a particular region over a set period of time, usually 30 years. The pattern is affected by the amount of rain or snowfall, average temperatures throughout the year, humidity, wind speeds and so on. Ireland has a temperate climate, in which it doesn't get too hot or too cold.

Climate change- A change in the climate of a region over time due to natural forces or human activity. In the context of the UN Framework Convention on Climate Change, it is the change in climate caused by higher levels of greenhouse gases in the atmosphere due to human activities as well as natural climate changes. See also global warming, and UN Framework Convention on Climate Change.

Conservation- Preserving or protecting animals and resources such as minerals, water and plants through planned action (such as breeding endangered species) or non-action

Deforestation- The reduction of trees in a wood or forest due to natural forces or human activity such as burning or logging.

Development - a developed tract of land (with houses or structures); the act, process or result of developing.

Development plan- A public plan that sets out the development objectives and policies of a local authority for its area.

Ecology - a branch of science concerned with the interrelationship of organisms and their environment.

Ecosystem - an interconnected and symbiotic grouping of animals, plants, fungi, and microorganisms.

Ecotourism- Small-scale tourism in fragile and protected areas that aims to have a low impact on the environment, benefit local communities and enable tourists to learn more about the natural and cultural history of the place.

Flora and fauna- The plants and animals that are native to a particular area or period of time.

Forests - lands on which trees are the principal plant life, usually conducive to wide biodiversity.

Global warming- The gradual increase in temperature of the Earth's surface caused by human activities that cause high levels of carbon dioxide and other gases to be released into the air.

Grazing - the use of grasses and other plants to feed wild or domestic herbivores such as deer, sheep and cows.

Greenhouse effect - the process that raises the temperature of air in the lower atmosphere due to heat trapped by greenhouse gases, such as carbon dioxide, methane, nitrous oxide, chlorofluorocarbons, and ozone.

Groundwater - water below the earth's surface; the source of water for wells and springs.

Habitat-The area occupied by a community or species (group of animals or plants), such as a forest floor, desert or sea shore.

Lakes - substantial inland bodies of standing water.

Over-development - expansion or development of land to the point of damage.

Over-grazing - grazing livestock to the point of damage to the land.

Population - the whole number of inhabitants in a country, region or area;

Reforestation- The process of planting trees in forest lands to replace those that have been cut down.

Renewable energy- Energy from renewable resources such as wind power, solar energy or biomass.

Renewable resource- A resource that can be used again and again without reducing its supply because it is constantly topped up, for example wind or sun rays.

Risk assessment - methods used to quantify risks to human health and the environment.

Surface water- Water that is collected on the ground or in a stream, river, lake, wetland or ocean.

Sustainable communities - communities capable of maintaining their present levels of growth without damaging effects.

Sustainable development- Development using land or energy sources in a way that meets the needs of people today without reducing the ability of future generations to meet their own needs.

Sustainable tourism- A form of tourism that meets the needs of current tourists and host communities while protecting and enhancing tourism for the future by balancing economic and social needs with a respect for different cultures and the environment. See also ecotourism.

1. Introduction

On 16th June 2013, in the middle of pilgrimage season, torrential rains and resulting devastating floods and landslides struck Uttarakhand and destroyed the Kedarnath town. Hundreds of people were feared killed and thousands of pilgrims were reported missing or stranded due to landslides, the state government is finally tackling the unprecedented disaster or natural calamity.

The Kedarnath valley, along with and other parts of the state of Uttarakhand, was hit with unprecedented flash floods on 16 and 17 June 2013 almost after 80 years. Aerial photos showed that the temple itself was still standing. However a portion of the temple complex was washed away, and the Kedarnath town was nearly destroyed in the deluge. The Kedarnath shrine would remain closed for a year for clearing the debris around the shrine. "It will take us at least five years to recover from the extensive damages caused to the entire infrastructure network in the Kedarnath area which is the worst affected," the Press Trust of India quoted him as saying. Dainik Bhaskar newspaper updated death toll (edition dated 25/06/2013), approach to 13000 death bodies. Hundreds have been recovered from under the debris so far and more could be buried. More than 60 bodies were found floating along the Ganga till bijnor city.

Kedarnath is a town located on the Garhwal Himalayan range near the bank of Mandakini river in Rudraprayag district, Uttarakhand, India. and has gained importance because of Kedarnath Temple a holy Hindu temple dedicated to Lord Shiva . It is located at 30°44′N 79°04′E. It has an average elevation of 3,553 metres (11,657 ft) above sea level near Chorabari Glacier, the head of river Mandakini, and is flanked by breathtaking snow-capped peaks.

The town suffered extensive destruction from flash floods caused by torrential rains. The temple complex and surrounding areas suffered extensive damage. However, aerial photos showed that the temple itself was still standing among the surrounding debris.

Land slide causing disruption of hill route, In view of importance of this landslide. It would "take a long time to rebuild Uttarakhand" and that no pilgrimage to Kedarnath would be possible "for at least the next three years".

In June 2013, flash floods and rain nearly destroyed the town of Kedarnath. The temple complex and surrounding areas suffered damage. thousands of pilgrims died in the natural calamity.

As the ecosystem of the Himalayan region is very fragile and delicate; change in one natural component goes a long way in transforming all other environmental elements, both biotic and abiotic.

Uttarakhand is known for great rivers like Ganga and glaciers like Gangotri, as per records the average rate of recession of Gangotri glacier is 20 meter per annum which is very alarming. A protest message against Tehri dam, which was steered by Sundarlal Bahuguna for years. It says- *We don't want the dam, The dam is the mountain's destruction.*

2. History

Kedarnath is the most remote of the Char Dham sites, Kedarnath is named in honor of King Kedar, who ruled in the Satya Yuga. He had a daughter named Vrinda who was a partial incarnation of Goddess Lakshmi. Lord Shiva is worshipped as Kedarnath, the 'Lord of Kedar Khand', the historical name of the region.

However, Kedarnath and its temple exist from the Mahabharata Era when the Pandavas are supposed to have pleased Lord Shiva by doing penance there.

Kedarnath is named in honor of King Kedar, who ruled in the Satya Yuga. He had a daughter named Vrinda who was a partial incarnation of Goddess Lakshmi. However, Kedarnath and its temple exist from the Mahabharata Era when the Pandavas are supposed to have pleased Lord Shiva by doing penance there.

3. Myth related to Kedarnath Temple

The Kedarnath Temple is one of the twelve Jyotirling, a holiest Hindu temples dedicated to the god Shiva and is located on the Garhwal Himalayan range near the Mandakini river in India. Due to extreme weather conditions, the temple is open only between the end of April (Akshay trutiya) to Kartik Purnima (the autumn full moon) every year.

It is believed that temple have been built by Pandavas and revived by Adi Sankaracharya. Pandavas were supposed to have pleased Shiva by doing penance in Kedarnath. Pandavas were supposed to have pleased Shiva by doing penance in Kedarnath.

The temple is also one of the four major sites in India's Chota Char Dham pilgrimage of Northern Himalayas.

The temple is not directly accessible by road and has to be reached by a 14 kilometres (8.7 mi) uphill trek from Gaurikund. The temple is not directly accessible by road and has to be reached by a 14 kilometres (8.7 mi) uphill trek from Gaurikund.

Temple is an impressive stone edifice of unknown date. According to Hindu mythology, during the war between the Kauravas and Pandavas, the kith and kin of the Pandavas were killed; in order to absolve themselves of this sin, the Pandavas undertook a pilgrimage. But Lord Vishweshwara was away in Kailasa in the Himalayas. On learning this, the Pandavas left Kashi. They reached the Himalayas via Hardwar. They saw Lord Shankara from a distance. But Lord Shankara hid from them. Then Dharmaraj said: "Oh, Lord, You have hidden yourself from our sight because we have sinned. But, we will seek You out somehow. Only after we take your Darshan would our sins be washed away. This place, where You have hidden Yourself will be known as Guptkashi and become a famous shrine."

From Guptakashi (Rudraprayag), the Pandavas went ahead till they reached Gaurikund in the Himalayas valleys. They wandered there in search of Lord Shankara. While doing so Nakul and Sahadev found a he-buffalo which was unique to look at.

Then Bheema went after the buffalo with his mace. The buffalo was clever and Bheema could not catch it. But Bheema managed to hit the buffalo with his mace. The buffalo had its face hidden in a crevice-in the earth. Bheema started to pull it by its tail. In this tug-of war, the face of the buffalo went straight to Nepal, leaving its hind part in Kedar. The face of the buffalo is Doleshwar Mahadev located in Sipatol, Bhaktapur, Nepal.

On this hind part of Mahesha, a JyotirLinga appeared and Lord Shankara appeared from this light. By getting a Darshan of Lord Shankar, the pandavas were absolved of their sins. The Lord told the Pandavas, "From now on, I will remain here as a triangular shaped JyotirLinga. By taking a Darshan of Kedarnath, devotees would attain piety". Near Kedarnath, there are many symbols of the Pandavas. Raja Pandu died at Pandukeshwar. The tribals here perform a dance called "Pandav Nritya". The mountain top where the Pandavas went to Swarga, is known as "Swargarohini". When Darmaraja was leaving for Swarga, one of his fingers fell on the earth. At that place, Dharmaraj installed a Shiva Linga, which is the size of the thumb. To gain Mashisharupa, Shankara and Bheema fought with maces. Bheema was struck with remorse. He started to massage Lord Shankara's body with ghee. In memory of this event, even today, this triangular Shiva JyotirLinga is massaged with ghee. Shankara is worshipped here in this manner. Water and Bel leaves are used for worship.

When Nar-Narayan went to Badrika village and started the worship of Parthiva, Shiva appeared before them. Nar-narayan wished that for the welfare of the humanity, Shiva should remain there in his original form. Granting their wish, in the snow-clad Himalayas, in a place called Kedar, Mahesha himself stayed there as a Jyoti. Here, He is known as Kedareshwara.

As a matter of fact, as one enters the main temple, the first hall contains statues of the five Pandava brothers, Lord Krishna, Nandi, the vehicle of Shiva andVirabhadra, one of the guards of Shiva. An unusual feature of the temple is the head of a man carved in the triangular stone fascia of the temple. Such a head is seen carved in another temple nearby constructed on the site where the marriage of Shiva and Parvati was held. Adi Shankara was believed to have revived this temple, along with Badrinath and other temples of Uttarakhand and he is believed to have perished at Kedaranath. Behind the temple is thesamādhi mandir of Adi Sankara.

The temple opens on Akshaya Tritiya (April end or first week of May) and closes on Bhai Duj (October end or 1st week of November) due to heavy snowfall and extreme cold weather during winter. Gaurikhund is 75 km (47 miles) from Rudraprayag. In between Rudraprayag and Kedarnath there are several places of pilgrimage such as Agastyamuni, Ukhimath, Phauli-Pasalat Devi Maa, Bamsu (Lamgoundi) Vanasur, Maa Jwalamukhi Devi village Andarwari, Maa Chandika Devi village lwara, Maa Kali at Kalimath, Trijugi Narayan (7 km (4 mi) from Son Prayag) (Where Lord Shankar got married to Goddess Parwati from when the fire of hawan kund is still alive) and Kashi Vishwanath at Guptakashi.

Udar Kund is located here. It is written in Kedar Khand of Shiv Mahapuran that the water of Udak Kund is mixture of all the 5 Oceans and always remain fresh even when kept for many years. The holy water of Udak Kund is used for purification rituals. Hans Kund, Bharo Nath, Navdurga Mandir, Shankaracharya Samadhi, Ishaneswar Mahadev Temple, Ret Kund, Panch Ganga Sangam, Chaurwari Taal now known as Gandhi Sarowar, Bashuki Taal are also places to visit here. There are several guest houses in Kedarnath with reasonable rates.

4. Uttarakhand: The land of God

Uttarakhand is comprised of 13 districts, namely, Chamoli, Pauri, Tehri, Uttarkashi, Dehradun, Haridwar and Rudraprayag in the Garhwal region and Nainital, Almora, Pithoragarh, Udham Singh Nagar, Champawat and Bageshwar in the Kumaon region. Of these 13 districts, four districts Nainital, Haridwar, Dehradun and Udham Singh Nagar have large areas in the plains, whereas the other nine districts comprise the hill region of the state.

Uttarakhand is a geographical inequality between the hills and the plains that divides the state critically. Districts in the plains are far ahead on various development indicators. The hill region districts are less developed in terms of infrastructure, i.e., electricity, roads and irrigation. The inter-district inequality in infrastructure leads to increasing disparity in terms of income and livelihood between the hills and the plains.

The vast natural resources add to the state's attractiveness as an investment destination, especially for tourism and agriculture- and forest-based industries.

Economy of Uttrakhand is absolutely dependent on mountains, are important sources of water, energy, minerals, forests and agricultural products and areas of recreation. They are storehouses of biological diversity, home to endangered species and an essential part of the global ecosystem. Therefore, the development of the mountains has to be viewed in a holistic manner, encompassing economic development, technological improvement, environmental protection and human resource development.

Himalayan terrain is highly prone to landslide hazards as compared to other physiographic divisions of India. Unstable geological conditions, heavy rainfall, cloudbursts, anthropogenic activities (excessive construction, deforestation etc.) etc. have been mainly responsible for several landslides resulting into loss of life and property or both. Due to these natural hazards, the Himalayan region is facing major problems of environmental degradation.

5. Causes of Environmental degradation in Uttarakhand

Geographical region of the Uttarakhand have been subjected to distinctive population pressures, over exploitation of biotic as well as abiotic resources, agricultural expansion etc. the Geologically, mountains in Himalaya are young have many fault zones and are subject to seismicity, denudation, landslide hazards and soil erosion. The highly seismic and unpredicted concentrated rainfall pattern are also responsible for landslide in Himalaya. Uttarakhand ecosystems like – forest ecosystem, agro-ecosystem, river ecosystem and soil ecosystem etc has degraded. On some places circumstances have reached at the very crucial stage like many streams and springs have dried up, forests have disappeared etc. These worst situation indicate administrations for the immediate action by human race to save the environment, from being inhospitable to living organism. Environmental disaster also pushes up poverty and emergence of so many types of problems etc. It is noticed that excessive run off losses of soil in denuded slopes, Hill areas and such other locations have driven the inhabitants to other areas in search of land.

These are some important causes of the degradation of various ecosystems of the region. Due to continues development activity the local ecosystems of the region is under great anthropogenic stress for the last several decades.

A few main causes are as follows :

1. Road construction activity

Construction of roads in Himalayan terrain is hazardous, if the existing instabilities are properly not accounted during excavations. Uttarakhand Government continues in the construction of new roads especially in erosion prone belt. It causes degradation of hydrological cycles and large scale down ward movement of the ruins to river valley. Uttarakhand is called as land of almighty. 85% revenue collected by pilgrimage/ tourism industry. For developmental activity need of expansion of road is essential, it is threat to the delicate ecosystem of this region.

Road construction activity involves cutting of forests and blasting of rocks this causes damage to the local eco-system.

2. Construction of Dams

Dam has been the object of protests by environmental organizations and local people of the region. In addition to the human rights concerns, the project has spurred concerns about the environmental consequences of locating a large dam in the fragile ecosystem of the Himalayan foothills.

There are further concerns regarding the dam's geological stability. Dam-break would submerge numerous towns downstream, whose populations total near half a million.

3. Expansion of agricultural land

In Himalayan region, the forest areas are converting into farm land in a macro scale. upslope extension of agricultural land has not only promoted the loss of biodiversity but it has also accelerated the pace of soil erosion and landslides. Expansion of agriculture is major requirement for sustain the local natives. Increasing population in hilly areas enforced the local people to convert their forest into agriculture land for cultivation.

The loss of top layer of soils is one of the prominent problems of the hills. Consequently Landslides constitute major natural disasters in the Himalayan ecosystem. The fragile Himalayan terrain often faces significant problems of geoenvironmental imbalance due to landslides.

4. Deforestation

Forestation in the hilly region is necessary for the formulation of ecological standards of the area through which an accurate assessment of the ecological situation. Rate of afforestation is much lower than the rate of deforestation. In last decade recurrence of fire in jungle is also major factor for the loss of all forest.

Deforestation is the major problems of the Uttarakhand. Forest cover protects the moisture, springs appear in the low lands and in ravine depths. But the areas which have been deforested and the green cover of the soil is eliminated, the soils have become dry which are unable to sustain crops. Gradually the uplands are going out of the agricultural use for dearth of water. It is known that the rainfall average is quite encouraging in hill areas but the water is drained to the per slopes which also washes away the fertile top soil and the grass-litter of the surface to the valley areas. Day by day Increasing agriculture, are now convert Uttarakhand as barren land owing to the loss of soil fertility through soil erosion. Gradually these forest lands are incapable of growing crops and growing natural grasses.

The density of forest tree is deteriorate in last decades. Forest fires destroy the leaf litters and seedlings on the ground, so that the humus contents of the soil reduce slowly.

Forest Development Ensuring adequate supply of fuel wood by installing massive energy plantations in, and around the villages can help the cause of meeting rural needs.

5. Grazing

Continuously growing requirements of milk and other dairy products, meat, wool etc. have always promoted livestock farming in the region. Excessive grazing and Livestock pressure on forest and lands is a powerful factor responsible for widespread degradation of forests in the region.

Large scale expansion of agricultural activities require more livestock population. On an average 3-5 hectares of forests are required to maintain a single cattle under the existing methods of grazing in the region, It has been estimated that only 40 per cent of the required grazing area is available in the central Himalaya . Unrestricted, open and free grazing of cattle may be checked by policy measures; and fodder banks may be established to cater the fodder needs of the farmers. Forest department must develop fodder supply zones and proper management method be used to regulate the grazing and consequent damages.

6. Development activities

Unemployment is the major problem for youth living on hill. Conventional occupation have sharply diminished and new jobs are almost not created by local government. Every year a large number of youth migrated to nearby cities on plains for employment.

There are several factors influencing the degradation of ecosystem, are fairly well known. Sumptuous bio diversity and other decisive environmental components of the Himalayan region being severely damaged as a result of increasing population pressure and adverse land use practices.

Such activities also accelerate the rate of erosion and therefore the instability of the slopes. Following this procedure ecological balance may be maintained to some extent.

7. How to conserve natural resources of Uttarakhand with some Important suggestions

Uttarkahand being a sensitive ecological zone, mountain environments are essential to the survival of the global eco-system. Mountains are, however, vulnerable to human and natural ecological imbalance. The Himalaya represent one of the most fragile mountain eco-systems and, furthermore, sustain a large human population. The terrain in the area is rugged and steep. development indicators must also reflect growth of environmental components especially forests, water, soil etc as major substances of common rural state. The status of these resources is not satisfactory but degrading largely.

1. Geology of the hilly area is important factor for estimation of groundwater resources, for construction of engineering structures and for abatement of landslides.

2. New roads should be constructed by covering mainly barren land and waste land. Forest land and agricultural land must be saved during the process of road construction.

3. Promotion of solar energy based apparatus might be fulfilled the need of energy on the hill sides, which is an extended exposure to sunshine (extra bright) due to the clean atmosphere. This energy can be well utilized for various purposes including pumping water from deeper rivers and water storage ponds.

4. Technological change is essential for raising efficiency, lowering the costs and reducing the strain. The appropriate technology development for farming operations in hill conditions has not been picked-up for research in proper scale. Hill areas have quite low width of the terraces and application of tractors, threshers etc. is not possible. The

threshing with bullock power is too strenuous and time-taking. Autoploughs of very small size partially operated by fuels and partially by human labour need to be developed.

5. Rainwater is hardly available and stream water is available only for the people living close to river-banks. there is need of proper rainwater harvesting structures in the region.

6. Revival of indigenous technology for rainwater harvesting and aquifer recharge, can ensure a dependable source of irrigation for crops. By this modern technology, a lot of water received from rain and seasonal springs can be stored for being used in summer towards use by livestock.

7. Gobar gas plants have been working successfully in the wider plains to cater to the needs of light and cooking. The most convenient and cost efficient method of obtaining energy in the rural areas is generating bio-gas.

8. Renewable energy sources like wind energy, bio energy (bio mass) can serve as a good cheap handy instrument for rural development in hill areas. Even the urban wastes which pollute the urban localities everywhere can be cost-effectively used for power-generation besides reducing environmental pollution.

9. The availability of cheap power by way of inexpensive water-lifting systems for irrigation will boost agriculture.

10. The village collaborated efforts can extract 'Shramdan' from the village youth if the social forestry is encouraged.

11. The Governments at the centre as well as in states today are increasingly alive to the need to reverse that continued process of deforestation which is the main cause of environmental degradation and has distorted the ecological balance of nature.

12. References

1. Wikipedia: the free encyclopedia

2. India floods: India floods: Death toll in Uttarakhand 'passes 500, BBC News. 21 June 2013. Retrieved 22 June 2013.

3. 4,000-year-old Mahabharata relic found in Nepal? (With Images). *The indian News*, 12 August 2009. Retrieved 7 July 2012.

4. Thapa, Bharat Bandu (6 Bhadra). "*Mandir Anabaran* ". Bhaktaput.

5. Prasai, Dirgha Raj. "Hindu shrine: Pashupatinath (Lord Shiva) and Shivaratri in Nepal". *The Indian Post*. Retrieved 7 July 2012.

6. Uttarakhand government website". Retrieved April 2007.

7. Monsoon fury leaves Kedarnath shrine submerged in mud and slush. *The Indian Express*. Jun 19, 2013.

8. Kedarnath shrine safe, to remain closed for a year. *The Hindu*. June 19, 2013.

9. A New Era of Economic Development – Uttaranchal: The Next Destination (2005).

10. Bisht, D.S. (2006). Poverty, Planning and Development - A Case Study of Uttaranchal State (submitted to the Planning Commission). Central Himalayan Institute. Dehradun. Trishul Publications, Dehradun.

11. Bisht, Sonali (2005). Concerns and challenges. In M.L. Dewan and Jagdish Bahadur (Eds.), Uttaranchal: Vision and Action Programme. Concept Publishing Company, New Delhi, pp. 82-87.

12. Mountain Research and Development, Vol. 18, No.3, pp. 213-233.

13. Dewan, M.L. and Jagdish Bhadur (Eds.) (2005). Uttaranchal: Vision and Action Programme. Concept Publishing Company, New Delhi.

14. Dewan, M.L. and K.D. Singh (2005). Forests resources: Regeneration and conservation. In M.L. Dewan and Jagdish Bahadur (Eds.), Uttaranchal: Vision and Action Programme. Concept Publishing Company, New Delhi, pp. 190-198.

15. Kar, Sabyasachi (2007). Inclusive Growth in Hilly Regions: Priorities for the Uttarakhand Economy. E/281/2007. IEG Working Paper, New-Delhi.

16. Kumar, Yogesh (2005). Micro-Hydropower Potential in Uttaranchal. In M.L. Dewan and Jagdish Bahadur (Eds.), Uttaranchal: Vision and Action Programme, Concept Publishing Company, New Delhi, pp. 102-109.

17. Malhotra, S.P. (2005). Opportunities, challenges and prospects in agriculture and forestry. In M.L. Dewan and Jagdish Bahadur (Eds.), Uttaranchal: Vision and Action Programme. Concept Publishing Company, New Delhi, pp. 54-66.

18. Mattoo, R.P., Anshu Singh, Dhananjay Kumar and Mudit Kumar (2004). Mechanism for Sustainable Development & Promotion of Herbal & Medicinal Plants in the State of Uttaranchal. Natural Resources India Foundation (NRIF) New Delhi (submitted to SER division, Planning Commission).

19. Monthly Review of Uttaranchal Economy (Oct-Nov-Dec 2006). Centre for Monitoring Indian Economy (CMIE), Dehradun branch office, Uttaranchal.

20. Pai, Rekha (2005). Evolution of Van Panchayats - Community participation in management and protection of forests. In M.L. Dewan and Jagdish Bahadur (Eds.), Uttaranchal: Vision and Action Programme. Concept Publishing Company, New Delhi, pp. 88-101.

21. Samant, P.K. and Palni, L.M.S. (2005). Resource conservation as a tool for economic development. In M.L. Dewan and Jagdish Bahadur (Eds.), Uttaranchal: Vision and Action Programme. Concept Publishing Company, New Delhi, pp. 242-262.

22. Sankhyiki Diary Uttarakhand (2005-06). Directorate of Economics and Statistics, Planning Department, Government of Uttarakhand, Dehradun.

23. Sharda, V.N. and Tayal, G.P. (2005). Potentials and problems of soil and water conservation in the development of Uttaranchal state. In M.L. Dewan and Jagdish Bahadur (Eds.), Uttaranchal: Vision and Action Programme. Concept Publishing Company, New Delhi, pp. 45-53.

24. Sreedhar, R. (2005). Mountain tourism for local community development. In M.L. Dewan and Jagdish Bahadur (Eds.), Uttaranchal: Vision and Action Programme. Concept Publishing Company, New Delhi, pp. 285-296.

Websites

1. http://www.organicuttarakhand.org/products_marketing.asp
2. http://www.yesbank.in/food_agribusiness.htm
3. http://www.projectsmonitor.com/detailnews.asp?newsid=10257
4. http://law.incometaxindia.gov.in/DitTaxmann/Notifications/IncomeTaxAct/2007/Notif251_2007.htm
5. http://www.sidcul.com/sidculweb/home.aspx
6. http://www.sidcul.com/sidculweb/inner_pages.aspx?cat_id=14
7. http://gov.ua.nic.in/
8. http://gov.ua.nic.in/uaglance/
9. http://www.hauraki-dc.govt.nz/documents/publications/EcoDev/Content.htm
10. http://gov.ua.nic.in/des/mainpage.htm
11. http://gov.ua.nic.in/AnnualDistPlan/
12. http://planningcommission.nic.in
13. http://districts.nic.in/
14. http://agricoop.nic.in/Agristatistics.htm
15. http://dacnet.nic.in/
16. http://www.uttara.in/
17. http://www.uttara.in/hindi/agriculture/intro.html
18. http://www.censusindia.gov.in
19. http://agcensus.nic.in/
20. http://www.rbi.org.in/home.aspx
21. http://www.ibef.org/aboutus.aspxChaturvedi, B. K. (2006), *Shiv Purana* (First ed.), New Delhi: Diamond Pocket Books (P) Ltd, ISBN 81-7182-721-7

ABOUT THE AUTHOR

Dr. Hemant Pathak held positions as Assistant Professor in the department of chemistry, Govt. Indira Gandhi Engineering College, Sagar, MP, India. He had extensive experience in teaching, research and administrative management.

Dr. Pathak received his Ph.D. degree in chemistry from Dr. Hari Singh Gour Central University, Sagar, India and M.Sc. Gold medalist from Jiwaji University, Gwalior. He has published 08 books and more than 50 research papers in reputed International and National journals and received several awards. He is a member of editorial boards and reviewer boards of several international journals and societies. His area of specialization includes Engineering Chemistry and Environmental Pollution management.